YOUR KNOWLEDGE HAS VALUE

- We will publish your bachelor's and
  master's thesis, essays and papers

- Your own eBook and book -
  sold worldwide in all relevant shops

- Earn money with each sale

Upload your text at www.GRIN.com
and publish for free

John Tarilanyo Afa

Statistical Analysis of Critical Flashover Voltage for small airgap

GRIN Verlag

Bibliografische Information der Deutschen Nationalbibliothek:

Die Deutsche Bibliothek verzeichnet diese Publikation in der Deutschen National-
bibliografie; detaillierte bibliografische Daten sind im Internet über http://dnb.d-
nb.de/ abrufbar.

Dieses Werk sowie alle darin enthaltenen einzelnen Beiträge und Abbildungen
sind urheberrechtlich geschützt. Jede Verwertung, die nicht ausdrücklich vom
Urheberrechtsschutz zugelassen ist, bedarf der vorherigen Zustimmung des Verla-
ges. Das gilt insbesondere für Vervielfältigungen, Bearbeitungen, Übersetzungen,
Mikroverfilmungen, Auswertungen durch Datenbanken und für die Einspeicherung
und Verarbeitung in elektronische Systeme. Alle Rechte, auch die des auszugsweisen
Nachdrucks, der fotomechanischen Wiedergabe (einschließlich Mikrokopie) sowie
der Auswertung durch Datenbanken oder ähnliche Einrichtungen, vorbehalten.

Imprint:

Copyright © 2011 GRIN Verlag GmbH
Druck und Bindung: Books on Demand GmbH, Norderstedt Germany
ISBN: 978-3-656-41177-2

This book at GRIN:

http://www.grin.com/en/e-book/213053/statistical-analysis-of-critical-flashover-
voltage-for-small-airgap

GRIN - Your knowledge has value

Der GRIN Verlag publiziert seit 1998 wissenschaftliche Arbeiten von Studenten, Hochschullehrern und anderen Akademikern als eBook und gedrucktes Buch. Die Verlagswebsite www.grin.com ist die ideale Plattform zur Veröffentlichung von Hausarbeiten, Abschlussarbeiten, wissenschaftlichen Aufsätzen, Dissertationen und Fachbüchern.

Visit us on the internet:

http://www.grin.com/

http://www.facebook.com/grincom

http://www.twitter.com/grin_com

JOHN TARILANYO AFA - UD67608HAE80826
SAC 672: Statistical Analysis of the CFO Voltage for small airgap

Atlantic International University

JOHN TARILANYO AFA

CURRICULUM DESIGN

TECHNICAL REPORT

SAC 672:

STATISTICAL ANALYSIS OF CRITICAL FLASHOVER VOLTAGE FOR SMALL AIRGAP

ATLANTIC INTERNATIONAL UNIVERSITY

HONOLULU, HAWAII

2011

Atlantic International University

TABLE OF CONTENT

Atlantic International University

CRITICAL FLASHOVER AND WITHSTAND VOLTAGE
(STATISTICAL APPROACH)

1.0   INTRODUCTION

Knowledge of electric fields is necessary in numerous high voltage applications. Electrode geometries that are encountered in practice or under laboratory experimental condition are very many but the rod-plane arrangement with rod tip of various shapes (cone, needle, sphere, co-axial cylinder etc) are more common (Onal, 2004). These gaps are subjected to several types of voltage shape so that breakdown and withstand voltage have to be ascertained.

Several models for explaining the observed characteristics of the fifty percent breakdown of long gaps have been put forward by many researchers (Begamudre, 2008). Some theories or models used experimental data to obtain numerical estimates for some factors. Even with the varying assumptions the results were similar because they were based on the same avalanche and streamer theories.

Based on the critical flashover (CFO) voltage of long gap put forward by Paris (Begamudre, 2008) there was need to relate this study to small gaps. This was particularly necessary since calculation of small gap field distribution is easier because the controlling factors could be checked.

The most important factors by which engineers describe the breakdown voltage of various gaps are the critical flashover and the withstand voltages. The critical flashover (CFO) voltage is the voltage that yields flashover of 50 percent of the number of shots given probability of voltage flashover ($V_{50\%}$).

The withstand voltage which is critical to the designer could be determined in various statistical form.

Due to large number of variables involved in insulation design especially the use of long gaps tends towards a statistical procedure because each of these variables has its own characteristics probability of occurrence either alone or in conjunction with other variables.

Also the probability of most events occurring simultaneously is also remote. Therefore a flashover once in many operations is allowed. The most usual case is to allow one flashover in 100 switching operations. This could form the basis for design of airgap insulation.

JOHN TARILANYO AFA - UD32863SFI20130
SAC 672: Statistical Analysis of CFO Voltage for small airgap

Some designers use 0.2% probability of flashover (1 in 500 operations). In insulation design and operation, the withstand voltage is of outmost importance. Thus, the low probability region of flashover, 0.1% or 0.2% must be obtained from flashover probabilities of higher values (Kara et al 2006, Faruk and Rizk 2009).

It is necessary to note that the flashover probability of a given gap distance with voltage follow a nearly Gaussian or normal distribution. The CFO and the withstand voltage could be related as

$$V_N = (1 - 3\sigma) V_{50\%} \qquad\qquad\qquad (1)$$

2.0 BACKGROUND OF THEORY:

Non-uniform fields: The important characteristic of non uniform field is the unequal distribution of field intensity in the space between the electrodes.

If the profile is different the greatest value of field intensity occurs on the surface of the electrode having smaller radius of curvature (Naidu and Kamaraju, 2007) and the region of minimum intensity is shifted to the opposite electrode. The degree of non-uniformity of a field can be characterized by the ratios of maximum intensity of the field E_m to its average value E_{av} that is,

$$K = \frac{E_m}{E_{av}} \qquad\qquad\qquad (2)$$

Non-uniformity of a field can exert considerable influence on the nature of the development of discharge (Afa and Wiri 2010, Marzinoto et al 2005)

In a sharply non-uniform field (positive point), an electron formed in the gap while moving towards the point falls into the region of strong field, ionization takes place near the point but at a distance sufficient to permit ionization and form avalanche. As the electrons of the avalanche reach the anode they move away through the point electrode but the positive charges due to their slow movement will remain in the space, moving slowly towards the cathode. This simulates an extension of the electrode surface (point) and its radius of curvature is increased thereby decreasing the field intensity in the vicinity of the point and slightly increasing the field on the cathode (Naidu and Kamaraju, 2007), Fofana and Beroual 2002).

If the voltage between the electrodes is sufficiently high an avalanche begins on the right hand side of the volume charge which by mixing with the positive ions of the volume charge creates

Atlantic International University

an embryo of the canal (anode streamers). The charges of the plasma of the streamer are situated in an electric field so that there are surplus positive charges. The charges partly compensate the field in the canal of the streamer itself and create increased intensity at its head. Presence of a region of strong field before the head ensures formation of new avalanches, the electrons of which are attracted in the canal of the streamer and ions establish the positive volume charges which lead to further increase of field before the head of the streamer. The newly formed avalanches convert this volume charges into the continuity of the canal of the streamer and this gradually spread to the cathode. The streamer growth for positive point electrode is shown in fig. 1.

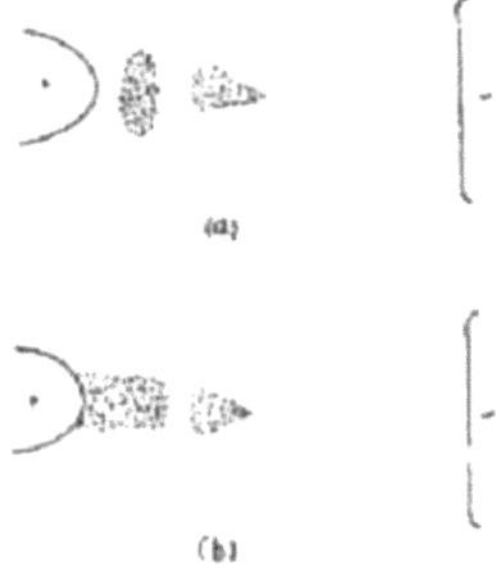

Fig. 1: Streamer Growth for Positive Point Electrode

Effect of humidity on breakdown voltage

Humidity in air exerts some significant influence on the breakdown voltage. According to data of Ritz (Razevig, 2007) an increase of absolute humidity of air from 10 to 25mm of Hg. column at normal atmospheric pressure of gap d = 1cm between electrodes increases the breakdown voltage by 2% in a uniform field. Kuffels investigation (Razevig, 2007, Fengnian 2003) for breakdown of air at normal atmospheric pressure showed that with smaller gap up to 20mm of Hg, the breakdown voltage increases from 4 – 5% in 85% relative humidity than in dry air. Figure 2 shows the effect of humidity on the breakdown voltage of a 25cm diameter spheres with spacing of 1cm when a.c and d.c voltages are applied. The spark over voltage increases with

partial pressure of water vapour in air, and or a given humidity condition, the charge in spark over voltage increase with the gap length.

The increase in breakdown voltage with increase of humidity is due to the fact that water vapour is electronegative gas. An increase of content of electronegative gas causes arrest of a large number of electrons with the formation of negative ions, as a result of which, the number of ionizing particles in the gap decreases and the breakdown voltage increases. However, in sharply non-uniform field the effect of humidity is considerably more. Usually, the effect of humidity of air is determined according to deviation from the breakdown voltage of standard humidity.

If the spark over voltage is V under test condition of temperature T and pressure P torr and if the spark over voltage is V_0 under standard condition of temperature:

$$V = kv_0 \text{-} \quad \text{-} \quad \text{-} \quad \text{-} \quad \text{-} \quad \text{-} \quad \text{-} \quad \text{-} \quad \text{-} \qquad (3)$$

Where k is a function of the air density factor d, which is given by:

$$d = \frac{P}{760}\left(\frac{293}{273+t}\right) = 0.386.\frac{P}{273+t}$$

The relationship between d and k is given in table 1.

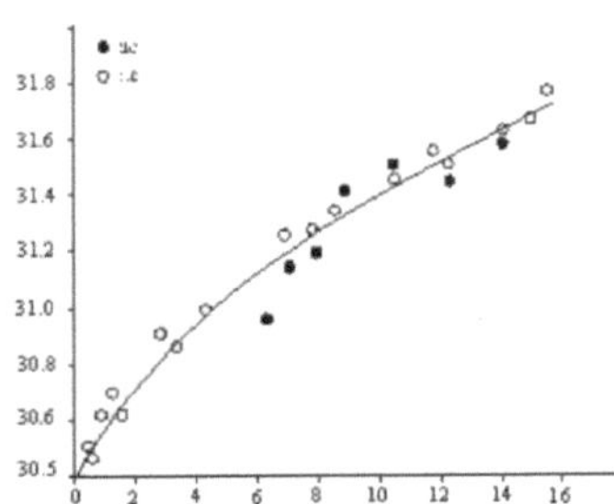

Fig 2: Humidity on breakdown Voltage for a.c and d.c for 25cm diameter spheres

Table 1: Relationship between correction factor k and air density factor d

D	0.70	0.75	0.80	0.85	0.90	0.95	1.0	1.05	1.10	1.15
K	0.72	0.77	0.82	0.86	0.91	0.95	1.0	1.05	1.09	1.12

The Stages of the High Voltage Equipment

The High Voltage equipment is made up of three stages, the A.C, D.C and Impulse stages. The circuit configuration is shown in fig. 3.

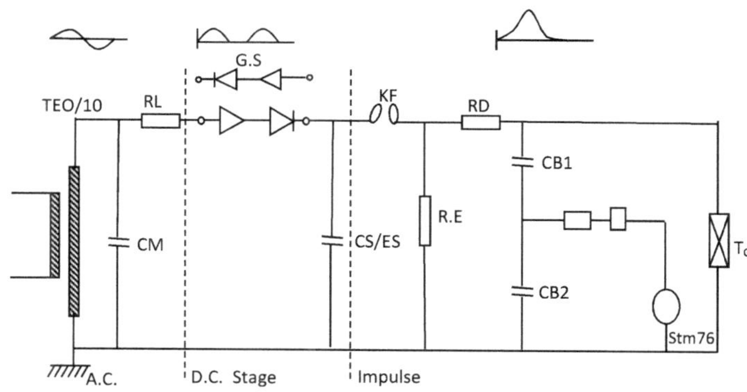

Fig. 3: Circuit Configuration of H.V. Equipment

2.1 A.C. Stage

The power frequency test voltage V was supplied by the test transformer (PTEO 100) and fed to the High Voltage measuring capacitor (CM). C.M is a capacitive divider for measuring power frequency voltage. The voltage if required is measured by tapping off the voltage at capacitor (CM), by the crest voltmeter SM76.

2.2 D.C. Stage

The voltage from the a.c. stage, charges the capacitor CS through the High Voltage rectifier GS (HV selenium rectifiers for supplying d.c. voltage). The load current flows to the measuring

resistor R.M through the protective resistor (Load Resistor). The d.c. output voltage if required is measured across the resistor R.M.

2.3   Impulse Stage

The spark gap assembly is responsible for the production of impulse voltage. It is supplied with a d.c. voltage of 140kv. It is a sphere-gap with the diameter 100mm and maximum gap distance of 80mm. A gap indicator moves along the graduated scale of the sphere gap with the operation of the drive for sphere gap assembly. The drive is made up of a split phase motor of 220v, 50HZ. The driver can be operated automatically from the control desk with the knob KF. The clearance between the spheres of the spark gap was set through insulating shaft.

After trigging the spark gap assembly KF, capacitor C.S. discharges to the capacitive voltage divider CB_1 and CB_2 and the test specimen through the damping resistor R.D. The impulse voltage is measured by the high resistance voltage StM 76.

The spark gap produces a standard wave. A standard wave has front time T_F of 1.2us with tolerance of 30% and the tail time T_F of 50us with tolerance of 10%.

If the voltage is high enough to produce breakdown on the sphere gap K_F, then the condenser CS (impulse capacitor) will discharge to the impulse circuit.

A voltage across the capacitor CB will vary according to the time constant T_F which is equal to the value of the product of the front capacitor and the wave front resistor R.D.

The time constant ($T_F = R_{FR} \times C_{FR}$) determines the time from minimum to maximum amplitude of the impulse. When the voltage across CB reaches its maximum, the capacitor CS which is parallel to CB is discharged completely to or through the wave tail resistor which determines the constant T_t of the tail time of the wave.

3.0 Materials and Methods

The impulse voltage was used for the breakdown test. The equipment (single stage H.V) operates up to 140kv on impulse voltage. The experimental set up is shown in fig 4.

Atlantic International University

Fig. 4: High Voltage Test Equipment

The electrode stand T_G was constructed from dried wood and ebonite pipes as shown in fig. 5. The wooden cross bar is made to carry a threaded iron rod of length 320mm where the electrode was fixed. The rod was also used for adjusting the airgap distance. The plate electrode (earth) is made of mild steel square plate (400mm x 400mm) that was folded and earthed.

JOHN TARILANYO AFA - UD12893SPO20139
SAC 672: Statistical Analysis of CFO Voltage for small airgap

Atlantic International University

Fig. 5: Electrode Stand (T_G)

A needle electrode of radius 0.25mm (tip) was used as the positive point electrode as shown in fig 6.

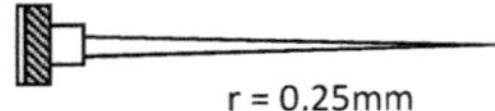

Fig. 6: Needle Electrode

With the help of the electrode stand T_G, the needle electrode was fixed and connected to the high voltage terminal of the high voltage equipment. The plate electrode serves as earth electrode. The gap distance of 10cm was maintained and the voltage and the spark gap K_f was adjusted to give a number of breakdown. This was repeated for four different voltages and the breakdown and probability was taken. For comparison the gap was adjusted to 5cm and 3cm and the probability of breakdown was also taken.

During the experiment the temperature, pressure, the wet and the dry mercury bulbs readings were taken.

JHH '"'" " "" " " " " " "
N M of 2, S " " "' Analysis " '" Voltage breakdown gap

4.0 RESULTS ANALYSIS

The results of the tests are given in table 2.

Table 2: Breakdown Voltage and Probability for Gap distances (3cm, 5cm and 10cm).

Gap distance cm	Breakdown Voltage (kv)	Breakdown Probability
3	40	2
	42	3
	46	6
	48	8
5cm	56	2
	58	3
	59	5
	63	7
10cm	78	0
	82	1
	86	6
	89	7

The most commonly used distribution function is the normal (Gaussian) distribution which has a particular bell shape. When the applied voltage V becomes the variable the Gaussian distribution function used takes the form

$$P(V) = \frac{1}{\sigma\sqrt{2\pi}}\, e^{-\left((V - V_{50})^2/2\sigma^2\right)} \quad - \quad - \quad - \quad - \quad - \quad - \quad (4)$$

Where, V_{50} is the voltage which leads to 50 percent probability of discharge and σ is the standard deviation.

A more convenient form of the normal distribution is the cumulative distribution function which has the form of

$$P(V) = \frac{1}{\sigma\sqrt{2\pi}} \int_{\infty}^{\infty} e^{-\left((V - V_{50})^2/2\sigma^2\right)} dx \quad - \quad - \quad - \quad - \quad - \quad (5)$$

A plot of the Gaussian cumulative distribution function is shown in fig. 7.

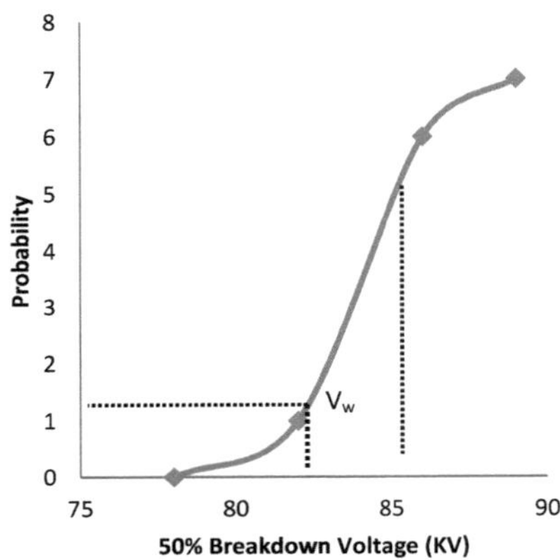

Fig. 7: Gaussian Cumulative Distribution Function

The graph of occurrence of overvoltage for 2% probability is shown in fig. 8

Fig. 8: Occurrence of Overvoltage probability

Atlantic International University

The experimental values (Breakdown Voltage) where plotted on a probability graph for 3cm, 5cm and 10cm and is shown in fig.9

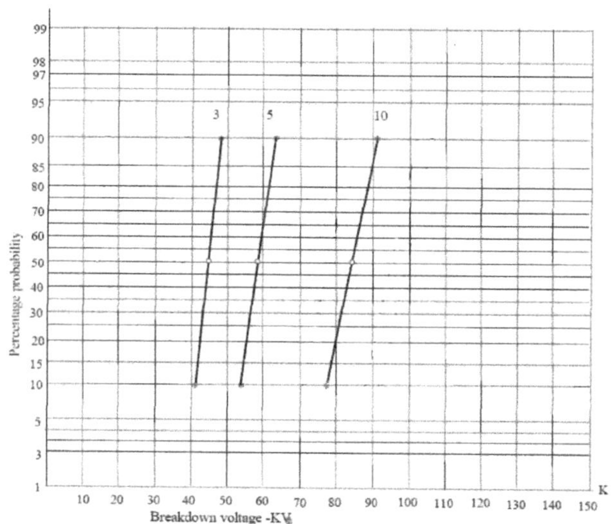

Fig 9: Breakdown Probability Graph

The distribution of flashover voltage of the gap is characterized by two parameters.

- V_{50} – called the critical flashover (CFO) or mean flashover voltage
- σ - called the standard deviation

The flashover voltage of the airgap expressed through the 50% flashover level (V_{50}) was read out from the graph of fig.9

$$V_{50\%} = 85kv$$

This is the mean or the critical flashover voltage (CFO) of the gap

The 10 percent and the 90 percent flashover voltage is given as

$$V_{10\%} = 78kv$$

$$V_{90\%} = 92kv$$

The standard deviation was calculated as

$$V_{50} - V_{10} = 85 - 78 = 7kv$$

$$= V_{90} - V_{50} = 92 - 85 = 7kv$$

The standard deviation $\sigma = 7kv$, and the mean is $85kv$

The coefficient of variation $= \dfrac{\%\,\sigma}{V_{50}} = \dfrac{7}{85} = 8.2\%$

$V_{50} \pm 1\sigma \ (KV)$		$V_{50} \pm 2\sigma \ (KV)$		$V_{50} \pm 3\sigma \ (KV)$	
$V_{50} - 1\sigma$	$V_{50} - 1\sigma$	$V_{50} - 2\sigma$	$V_{50} - 2\sigma$	$V_{50} - 3\sigma$	$V_{50} - 3\sigma$
78	92	71	99	64	106
33%	66%	2%	95%	0.13%	99.87%

From the normal (Gaussian) distribution curve, the withstand voltage was estimated. The voltage range over which the probability of flashover (CFO) was $\pm 3\sigma$

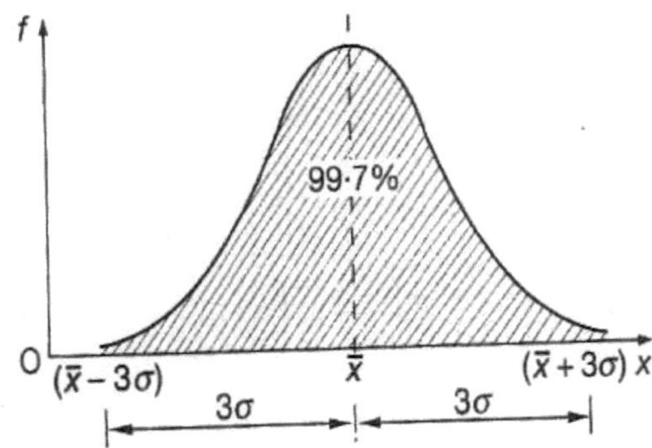

Fig. 10: Values within 3sd of the mean

1. $CFO - 3\sigma$ is known as the statistical withstand voltage and represents the point with flashover probability 0.13 percent.

AIU UGREEANTE D A... ... ...
... ... Statistical Analysis of a EHV voltage for small air gap

2. $CFO + 3\sigma$ is known as the statistical flashover voltage (SFOV) and represents the point with flashover probability of 99.87 percent.

5.0 DISCUSSION

The experiment was carried out in February and the relative humidity was low, the pressure was almost equal to the normal atmospheric pressure, therefore, the correction factor has little effect on the flashover voltage.

Estimating the critical flashover voltage (CFO) using the statistical method (probability graph) gave a very close approximation to the values given by the formular for breakdown for small gap (Afa and Wiri 2011).

In engineering practice it would become uneconomical to use the complete distribution function for the occurrence of over voltage and for the withstand voltage of the insulation.

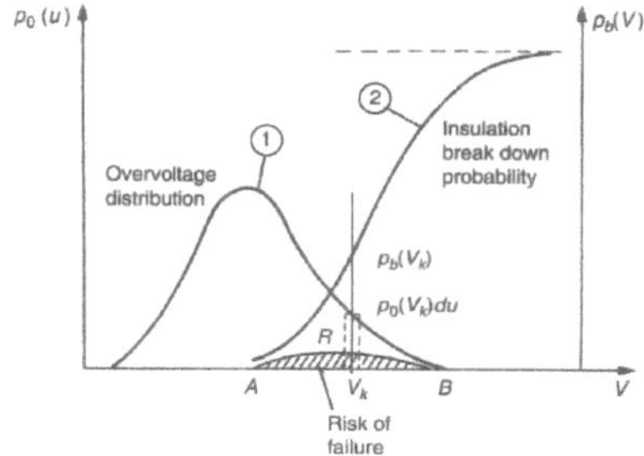

Fig. 11: Method of describing the risk failure and safety factor

Fig.8 represents probability of occurrence of over voltage of such amplitude V_s that only 2 percent (shaded area) has a chance to breakdown. In fig.7 the impulse breakdown occurs at 10 percent, the voltage V_w is known as the statistical withstand voltage. When fig.7 and fig.8 were plotted together as shown in fig.11, the failure risk could be determined from the figure (fig 11).The failure risk f_r

$$f_r = \int_o^\infty P_b\,(V_k)\,P_o\,(V_h)\,du - \quad - \quad - \quad - \quad - \quad - \quad (6)$$

Where

$P_b\ (V_k)$ = Probability of breakdown

$P_o\ (V_k)du$ = Probability of occurrence of over voltage

V_k = the Kth values of over voltage.

6.0 RECOMMENDATIONS

❖ Due to the variation of breakdown voltage as a result of climatic change, it is necessary that the humidity, Pressure and Temperature must be normalized in the calculation for accurate results. That is, correction factor is necessary.

❖ Due to the gap distance, withstand voltage was worked out with lightning impulse voltage.

7.0 CONCLUSION

Many researches have been done on long gaps because most practical situations are based on long gaps. The work was necessary because in small gaps the calculation of field distribution will lead to a better understanding of breakdown characteristics because the secondary phenomena could easily be controlled. Once these mechanisms are thoroughly understood, the breakdown characteristics of very long air gaps have been known to follow the same pattern as small gaps.

The gap under test has a breakdown voltage between 80kv to 110kv therefore test was limited to standard lightning impulse test with a voltage wave of 1.2 / 50 µsec. For much higher voltage wave switching impulse voltage would have been included.

The statistical approach to insulation presented in the work, leads to finding out of withstand voltages of the small airgaps that is, the probability of breakdown for small gap could be established.

The statistical approach of insulation co-ordination relates directly the electric stress and electric strength which requires the knowledge of distribution of both the anticipated stresses and the electrical strength.

Atlantic International University

For the purpose of co-ordinating the electric stresses with electric strength it was convenient to represent the breakdown voltages distribution in form of probability density function (Gaussian distribution curve) and the insulation breakdown probability by the cumulative distribution function.

The combination of these distribution functions became a viable tool for finding the risk of failure and the statistical safety factor for the insulation.

It is known that increasing the statistical safety factor (S.F) reduces the risk of failure but at the same time will increase in insulation costs. This will enable designers choose the level of statistical safety factor in terms of the type of insulation required (self restoring insulation etc).

8.0 REFERENCES

[1] Afa, J.T., 2010: Comparative Analysis of Breakdown Characteristics for small and long Airgaps: European Journal of Scientific Research 45(1) 324 – 332.

[2] Afa, J.T. and Wiri, A.T., 2010: Exponential Expression of Relating Different Positive Point Electrode for small airgap Distance: Research Journal of Applied Science, Engineering and Technology, 2(6) 512 – 518.

[3] Begamudre, R.D., Extra High Voltage A.C. Transmission Engineering 3rd Ed.: New Age Int. Publishers.

[4] Faruk, A.M. and Rizk, 2009: Modelling of Proximity Effect on Positive Leader Inception and Breakdown of Long Air Gaps: IEEE Trans. On Power Delivery, 24(4) 2311 – 2319.

[5] Fengnian, H., 2003: A New Statistical and Optimum to Lightning Insulation Co-ordination of Subsation: Electrical Power Research 23 (2) 129 – 137.

[6] Fofana, I. and Beroual, A., 2002: A model for long airgap discharge using an equivalent electrical network: IEEE Trans. Of Dielectrics and Electrical Insulations, 3(2) 273 – 282.

[7] Kara, A., Kalenderli, O., and Mardiyan, 2006: effect of Dielectric Barriers to the Electric Field of Rod-Plane Air Gap in Pro. Of the COMSOL, Users Conference, Prague.

[8] Marzinotto, M., Mazetti, C., and Schiaffino, P., 2005: Statistical Approach to the Insulation Co-ordination of Medium and High Voltage Cable lines: Power Tech, IEEE Russia.

[9] Naidu, M.S. and Kamaraju, V., 2007: High Voltage Engineering. Tata McGraw-Hill Publishing Company Ltd., new Delhi.

[10] Onal, E., 2004: Breakdown Characteristics of Gases in Non-Uniform Field. Journal of Electrical and Electronics Engrg. 4(2) 6022 – 615.

[11] Razevig, D.V., 2008: High Voltage Engineering, Khana Publishers, New Delhi, India.

[12] Shunguang, G.U., Chen, W., Chen, J., He-Hengxin Quan, G., 2010: Observation of the Streamer Leader Propagation Processes of long airgap Positive Discharges: IEEE Trans. On Plasma Science, 38(2) 214 – 217.

JOHN JORDAN OSAMA UTHMAN ISPO PGS D
SAC 8822: Statistical Analysis of CFO Voltage for small airgap

Atlantic International University

[13] Wadhwa, C.L., 2008: High Voltage Engineering: New Age International Publishers, New Delhi.